Human Genetics
Its Influence on Behavior and the Immune System

PART A: BEHAVIOR

Ronnee Yashon

NEWMAN SPRINGS PUBLISHING
320 Broad Street
Red Bank, NJ 07701

First originally published by Newman Springs Publishing 2022

ISBN 978-1-68498-369-8 (Paperback)
ISBN 978-1-68498-370-4 (Digital)

Printed in the United States of America

For everyone who inspired me along life's journey.

Contents

Section A

Behavior

Preface

This book will look at behavior in a different way. Have you heard of the phrase *nature versus nurture*? Simply, it asks what influences our behavior. This has been a debated topic since early man. *Nature* is usually defined as what is given to us before we are born, specifically, in this book, genetics. *Nurture* means learning that is acquired from our environment (parents, friends, and other influences). In addition, we will look at the newest scientific work—how both genetics and environment affect how we behave. Our court system is slowly changing as science rears its head in the area of criminal law, everyday life, public opinion, and politics.

Table P.1. Are these nature or nurture? Or both?

Right- or left-handedness?	Even though this is known to be genetic, it can be changed by training.	A number of years ago, teachers were told not to allow lefties to use their dominant hand.
Violent personality?	Brain chemicals and hormones control this.	A tie-in to violence has been influenced by brain chemicals.
Athletic prowess?	Certain traits (height, muscularity) can be enhanced by drugs and other stimulants.	Can someone short be a great basketball player?
Color blindness?	Nature	Always genetic or cannot be cured or changed, except for contact lenses.

Sleepwalking?	No	Usually triggered by stress.
Intelligence?	There are theories (eugenics).	Robert Klark Graham began a sperm bank he called the repository for germinal choice, also known as the genius sperm bank.

CHAPTER 1

Behavior

Simply, behavior can be defined as a reaction to a stimulus; this sounds pretty silly because if we look at one stimulus, it can have many reactions among many people. The stimulus might be a loud sound or a complicated thought.

Scientists studied the behavior of animals and wondered why they do certain actions. One of the early animal behaviorists (Pavlov) studied what dogs can learn. Since then, animal behavior is still studied but with an eye to comparing it with humans.

If you watch an animal (say a dog), you will see it learns new behaviors if rewarded (usually with food). But then, so do humans; advertising is an example.

Does animal behavior predict what behavior humans might have?

The answer is *probably*. So why are we testing on animals first?

Because we do not want to test on people, possibly because of any number of consequences, such as side effects, lawsuits, mistakes, bad publicity, and of course, loss of revenue.

But if you look at animals versus humans in drug testing, you can see how dangerous it can be for drug companies to skip the animal trial if something would go wrong when humans are using it.

Note: Why this is important in courts?

As humans, we feel in control of our actions, so if we lose control, we might feel angry, guilty, sorry, or happy.

CHAPTER 2

Physiology 1

In humans, behavior is controlled by the brain and spinal cord. Reflexes are a perfect example. They react quickly and protect us from more harm.

Our first behavioral multiple-choice question 1

You walk into the kitchen, and a pot is boiling over on the stove. What do you do?

- a. *Ask someone.*
- b. *Look for clues (burner on, smell of food).*
- c. *Touch the pot to see if it is hot.*
- d. *Walk away.*

Answer. All of these might be correct, and your response may change because of your age and previous knowledge. Most of our behaviors begin with a stimulus traveling to the brain *first*, but not reflexes. Your reflexes tell your body, *danger*! The stimulus then goes to the spinal cord, which immediately sends a message to the arm muscle to contract and pull your hand away. It bypasses the brain and solves the immediate problem. Because nothing is perfect, you might still get a burn, but not as severe. Our brain will store this reflex. See illustration 2.1.

There are two types of reflexes: the autonomic reflex, affecting inner organs, and the somatic reflex, affecting muscles.

Other reflexes: blinking, pupil opening and closing (in response to light), arm or leg responding to a pinprick (in response to pain), coughing or sneezing (response to irritants or allergens in the nasal cavity), and knee jerk (response to a blow to the knee).

Other brain functions are routed through different areas of the brain (see illustration 2.3). Sight is a good example. When light enters the eye, it is turned into electrical impulses by passing through the retina. These impulses travel up the optic nerve into the spinal column and up to the brain.

If you choose c, you might be hurt, or your reflexes might kick in (see illustration 2.1).

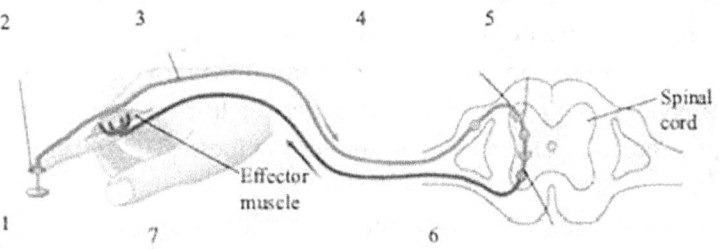

Illustration 2.1. Reflex

Reflex arch: A reflex arc is a neural pathway that controls a reflex. In vertebrates, most sensory nerves do not pass directly into the brain, but synapse in the spinal cord. Follow the steps in illustration 2.1.

1 and 2: Stimulus (heat, injury, or pain)
3 and 4: Carrying the impulse to the spinal cord
5 and 6: Impulse enters the spinal cord and immediately leaves, carrying the impulse to the muscle
7: Muscle that is stimulated moves the finger away

In our example, the impulses are sent to the area of the brain that controls sight (see illustration 2.2) and the impulses are read. Then, they are converted back into impulses and sent to the eye, and this is what our brain identifies as an image.

Think about how long it takes to notice something, look at it, and, identify it. It happens almost immediately, and you see what your eyes were sending to the brain.

You see it, and then, you immediately remember you have seen it before.

What if it is something you have never seen before? Your brain might identify it as a familiar sight. But if you do not identify it, you will be able to find something close to it, and your brain takes over to give you the answer.

This is just one of the amazing things that the brain does: It also stores information, makes information available to you, and controls your emotions and behaviors. *All in a split second.*

Many things can go wrong with impulses moving through the peripheral nerves to the brain. If a person loses a neural connection, the optic (eye) center does not get the message and could cause blindness.

Examples: If a tumor on the optic nerve stops the impulses, a serious spinal cord injury could cause paralysis of the lower half of the body.

Scientists usually separate behaviors into innate behaviors (present at birth) and learned behaviors (behavior that is acquired through experience, observation, and teaching).

These two types of behaviors can also be called *nature versus nurture* (see Chapter 1). In this book, we will be looking at both sides of this argument.

Is this *really* an argument?

Short example: Much of what we will cover in this book is determining which behavior leads to action and is this action reasonable. If you are mad at someone, you might slap them. What causes this behavior? Was it automatic? On purpose? Because of something he or she said or did?

The legal system tries to understand motivation but sometimes cannot clearly define it.

Example: Premeditation? An accident? Or insanity?

Behavioral multiple-choice question 2

You are on a jury, the crime is murder, and the suspect says he was sleepwalking and was not responsible. How do you vote?

a. *Not guilty, because he could not control what he did.*
b. *Guilty, because sleepwalking is not a defense.*
c. *Not guilty, because he was not aware of what he was doing.*
d. *Trial is suspended because the judge does not understand the evidence. (Science?)*

CHAPTER 3

Physiology 2

The areas of the central nervous system (CNS) that control our behavior are the *brain* and the *spinal cord*.

Second, the peripheral nervous system (PNS) is composed of all nerves that are moving impulses from the CNS to different areas of the body. (This is how pain is sent. In our example, impulses move to a certain area to alert you to a problem with pain.)

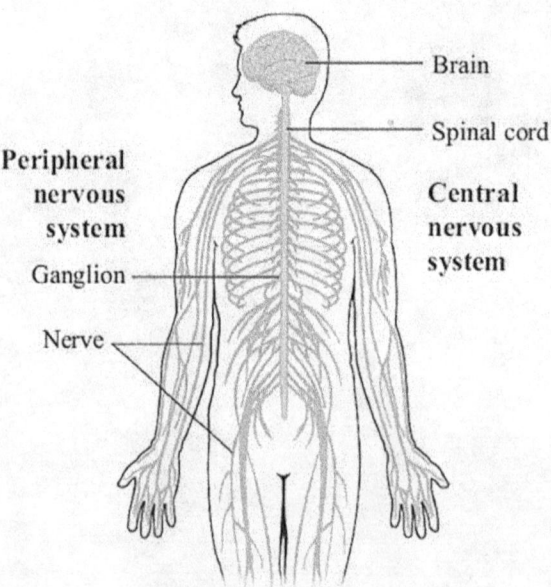

Illustration 3.1. The PNS

Lastly, a fast response (reflex, see Chapter 2) is caused when an action must be acted on (see illustration 3.1) immediately. In an emergency, we do not even have time to study what is happening to us. In our example, the muscle would react very quickly to keep us from burning.

The body might respond with various actions. Some of these are sweating, heavy breathing, and running; and all or some of our thoughts are often thought of as *crazy*.

Note. If you are taught as a child not to take food from strangers, it will become a permanent reflex over time.

Now, think about criminal activity as an example. Our laws are there to give us an idea of what is right or wrong. But what is right or wrong depends on the situation. Law has been asking the question, Why did someone commit a crime?

Remember nature versus nurture (see Chapter 1)?

From serious crimes to taking an extra apple, all these involve the nervous system.

Take, for example, a man who is on a trial for murder. Is he innocent or guilty? What should be his punishment? If he is in an accident that caused brain damage as a child, does this play a part?

Defense argument. Brain damage made it impossible to make good decisions. He might be found not guilty even though he did commit murder.

This has become one of the big problems in criminal trials for many years. We will look into this later.

CHAPTER 4

Neurotransmitters

Neurotransmitters are chemical molecules synthesized within the brain cells. They allow the transfer of signaling messages between the brain cells. The movement from nerve ending to nerve ending is controlled at the synapse (see illustration 4.1).

What do the brain chemicals do?

They are also called neurotransmitters; this word tells you all you need to know.

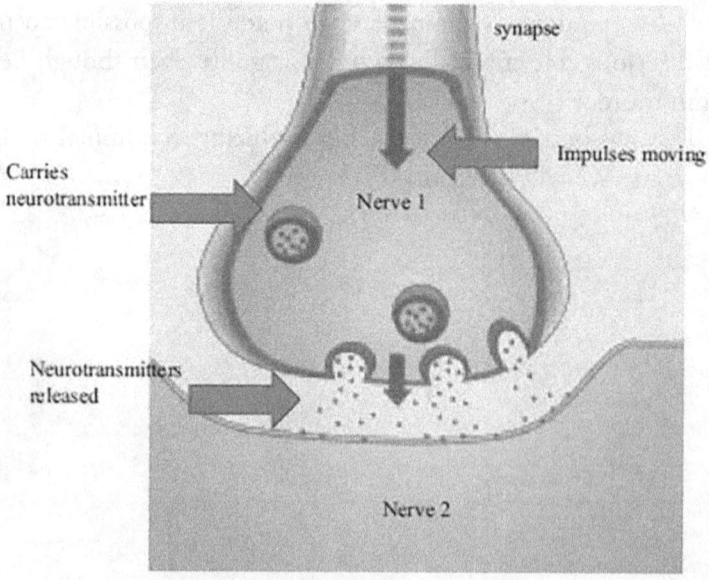

Illustration 4.1 The impulses move through the synapse

Just separate the parts of the words *neuro* (nerves) and *transmitters* (help the transition), and you can see that chemicals help impulses travel to the next neuron (see illustration 4.1).

Neurotransmitters all serve different purposes in the brain and body. There are many of these, but we will pick out the major six: acetylcholine, dopamine, norepinephrine, serotonin, gamma-aminobutyric acid (more commonly referred to as GABA), and glutamate (see Table 4.1).

Question. What are the effects of the release of these neurotransmitters? (see Table 4.1)

Table 4.1 Major neurotransmitters

Name	Purpose	Effect
Acetylcholine	Found throughout the nervous system. It is the only neurotransmitter that sends and receives information between the motor neurons and voluntary muscles (muscles you have conscious control over, such as the biceps).	Muscle stimulation Heart rate increase Sleep (avoid)
Dopamine	Regulates many aspects of behavior, including pleasure, emotion, and behavior (movements).	Learning Memory Mood Positive thinking
Norepinephrine	Also regulates behavior.	During stress, it ups the heart rate and courses the blood away from the digestive systems to the muscles. Called *fight or flight*.
Serotonin	Regulates mood and plays a major role in sleep, wakefulness, and eating.	Sleep Mood Appetite Pain Body temperature

GABA[1]* and glutamate	During stress, it helps us stay calm and not overreact.	Control during stressful situations (many be fixed with drugs).
Endorphins (three types)	When released, often during exercise, a rush of excitement and happy feelings result.	

*Gabapentin

These neurotransmitters affect our behaviors and feelings (one example is a *runner's high* when exercise leaves you feeling happy).

What is the effect of slow or no releases of neurotransmitters? Misconnection at the synapse, which makes it impossible for impulses to go on. Mutations in the genes that produce neurotransmitters are called synaptic malfunctions and have the same effect.

Many pharmaceuticals (prescription or street drugs) mimic neurotransmitters and have similar effects (e.g., cocaine and nicotine mimic endorphins).

Do you think pharmaceutical companies knew about this?

In addition to neurotransmitters, our brain is affected by the release of hormones from the sex organs; these hormones rise and fall during certain times.

Hormone	Whom does it affect?	Increase in hormones occurs naturally	When hormones are released
Testosterone	Men**	Sex, anger, puberty, steroids	Violence, strength, may cause rage (steroids)
Estrogen*	Women***	Puberty, menstruation, and pregnancy	Pregnancy and birth

The two hormones in question here are estrogen (female) and testosterone (male).

**The major hormone in men is testosterone, but men have estrogen too. High estrogen in males can cause infertility, erectile dysfunction, and gynecomastia (enlarged breasts).

***Women have both estrogen and testosterone. When the ovary cannot make estrogen (menopause), the estrogen stops, and testosterone begins to become active. Symptoms might be facial hair growth and baldness, and other male secondary sex characteristics may show up.

*Estrogen increases when a woman is pregnant because the hormone is needed to keep the embryo in the uterus. When birth begins, estrogen drops significantly, and the uterus expels its contents. This includes the fetus. If it is the time of birth, contractions begin. After birth, estrogen drops quickly and may cause deep depression (postpartum depression) or, worse, postpartum psychosis. Some mothers have killed their children (see Chapter 10).

CHAPTER 5

How Might Genetics Change Behavior?

In a number of ways, the following are examples of how genetics might change behavior:

1. Chromosomal aberration (extra Y). See Table 5.1.
2. Single-gene defects (caused by dominant or recessive genes). See Chapter 6.
3. Brain damage from injury or birth. See Table 5.1.
4. Epigenetic changes in the DNA itself during pregnancy caused by pollution and other environmental changes.

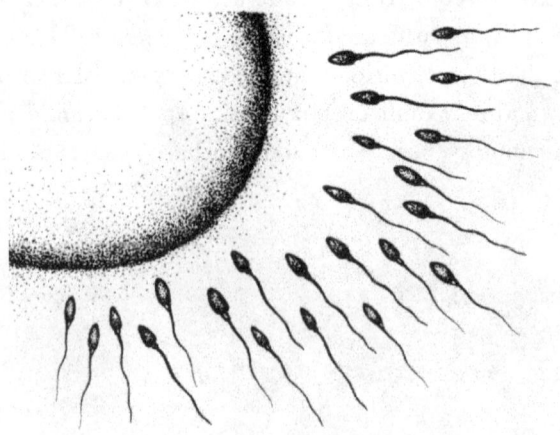

Illustration 5.1. Sperm

Table 5.1. Genetic conditions that cause changes in behavior

Condition	How does this affect behavior	Symptoms	Diagnosis	Test	Treatment
Brain damage either before or after birth	Any problem during pregnancy or birth that causes brain damage.	Slow reflexes, cognitive skills later in life, deafness	Yes, before birth (amniocentesis) or in the first few months	Ultrasound before birth showing smaller brain size or a brain damage	Physical therapy
Fragile X (broken X chromosome)	This happens during meiosis in the sperm or egg.	Mental retardation and explosive reactions	By taking cells from the fetus and testing a cell (amniocentesis)	Test for extra or a broken chromosome (amniocentesis)	Treat as they occur
Schizophrenia (inherited mental illness)	Shows up in late teens.	Behavioral problems, hallucinations, depression	Psychiatric help, institutional commitment	Diagnosis by analyzing behavior	Medication
Type 1 and 2 diabetes	Extremes of blood sugar can affect behavior.	Dizziness, coma, hallucination, and if not corrected immediately, death	Control of blood sugar with insulin injections or insulin pump	Type 1 is genetic, and a child is usually diagnosed (very difficult to control). Type 2 occurs usually in middle age and can be controlled with medications, diet, or insulin	Insulin, diet, and testing of blood
Chromosomal aberrations (see explanation next)*	XXY XYY	Anger, violent behavior, hyperactivity, learning	Amniocentesis and karyotyping. Can be done *in utero* and the diagnosis is given to the parents. They decide what action should be done.	No symptoms would show up at birth. This can be determined from a chromosome.	Separate sperm and delete only Y bearing into the egg

*Think back to how a fetus's sex is determined.

Remember this?

X and Y are the two sex chromosomes.

Where do they come from? They formed in the ovary and testis cells. All eggs have an X. Sperms have *either* an X or Y.

Notice. This configuration means that when egg and sperm join, the egg *always* contributes an X. So if you think about it—it is the male sperm that determines the sex. (If an X-bearing sperm joins the egg—*it is a girl!* If a Y-bearing sperm joins the egg—*it is a boy!*)

Sounds simple, right? (See next.) But actually, *what* controls which sperm joins the egg?

If the couple wants a boy, is there anything they can do to make that happen?

How does an XYY configuration occur?

The Y chromosome creates the male hormone (testosterone) if two *Y*s exist in a fertilized egg; it seems obvious it might affect the testosterone levels in the fetus and the adult as well. What causes this (XYY) to happen?

Actually, scientists are not sure; some have theorized (see Table 5.2).

Table 5.2. Theories and Studies of XYY formation

Theory 1	Two sperm join the egg	Yes, but under most circumstances, this would create a zygote with 69 chromosomes.*
Theory 2	Uneven meiosis, causing chromosomes to stick together in sperm	During the process, chromosome pairs must separate.
Theory 3	Separation of Y chromosomes during meiosis	During the process, chromosome pairs must separate.
Theory 4	Sperm weight Lighter sperm swim faster? True?	Y-bearing sperms have a chromosome (Y) that is half the size of X.
Theory 5	The father of this child also has an extra X in some sperm	Both the grandfather and father carry the extra Y. This can be determined by taking a sample from each.

Only embryos that carry the correct number will live.

**Most of these studied XYY men are in prison for violent crimes.

Year	Who studied?	Where was it done?	
1962	A male who had a Down syndrome child	Denmark	First patient tested
1966	Richard Speck Mass murderer (see in a murder case in Chapter 10)	First time brought up in a murder case	The testing was approved, but Speck did not have the extra Y.*
1970s		Patients in mental institutions convicted of serious crimes The thinking was, XYY meant criminality.	The feeling proved untrue.
1978–2006	Denmark	Long-term study did not involve any change in criminality.	The risk of future conviction for those with XYY was slightly elevated.

*Speck's trial was covered heavily in the press. Many thought if he tested positive for XYY, it would prove his guilt, but he was not XYY, and he was found guilty because of a witness.

Many studies have been done and shown, but many still believe an extra Y means more testosterone and, therefore, more aggression.

Think about this: In the far future, all newborns will have to have their complete DNA sequence done. If the child's karyotype showed XYY, what might be done with this child?

CHAPTER 6

Does Any Single-Gene Defect (Mutation) Cause Aggressiveness?

Could this be used as a defense in court?

What is monoamine oxidase A (MAOA) deficiency? Monoamine oxidase breaks down neurotransmitters in the synapse, so when its gene is mutated, the next impulse would not continue through the synapse. This might cause changes in the behavior. Its gene is on the X chromosome; see illustration 6.1.

Illustration 6.1

This case is an interesting example of genetic testing and how it can affect a single case (and cases to come). Think about this as you read the following case: If the *cause* of Mobley's action and reaction was controlled by his mutated gene for MAOA deficiency, should his verdict reflect this?

Mobley v. Georgia
455 S.E. 2nd 61 (1995)

Facts of the case

On February 17, 1991, John Collins was shot in the back of the head. This occurred during a robbery of the Domino Pizza Store in Oakwood, California.

Mobley worked in Oakwood, California, franchise. On March 13, after being picked up for questioning, Stephen Mobley confessed to the murder and armed robbery.

During this confession, he boasted how John Collins fell on his knees and begged for mercy. Then, he had himself tattooed with a Domino's logo and plastered his cell with Domino's boxes. His history included rape, robbery, assault, and burglary. The prosecution said, "Mobley is evil, a cold-blooded, heartless killer."

Daniel Summer, the court-appointed attorney for Mobley, tried to get a plea of guilty entered and a deal for life in prison, but the deal was rejected. The court was interested in the death penalty for Mobley.

During the questioning of Mobley's family, Summer met Joyce Ann Childers, his aunt. She told Summer that "volcanic, aggressive, physical abuse and violent behavior is prevalent throughout the family tree." Summer then remembered an article he had read in *The Chicago Tribune* in which scientists at Harvard, the National Institutes of Health (NIH), and labs overseas were conducting research on genetic ties to violence.

Before the sentencing hearing, Summer contacted the doctor from Harvard Medical School and another expert at Emory University.** When Mobley's story was revealed, the doctors began to see a pattern emerge of a family history of violence moving through a number of generations in Mobley's family (see illustration 6.1). Both doctors offered their services free of charge, but Mobley would need specialized testing of urine and blood to determine if he suffers from a genetic mutation of the gene for MAOA similar to the patients studied by Breakefield and Farat.*

One might expect Mobley to come from a poor family and bad surroundings; the opposite is actually true. Even though many of the people in Mobley's family tree were violent, many were amazingly successful. His father, for example, even though he refused to help Mobley's defense, is a self-made millionaire. He tried sending his son to private school, then psychiatrists, and finally, jail but stated, "He never developed a value system or a conscience." In the end, he washed his hands of the responsibility.

The cost of the tests, about one thousand US dollars, was not available to Mobley, and Summer asked the court to release funds for the test. This allowed the trial court and, eventually, the Supreme Court of Georgia to weigh the question of the validity of genetic causes of criminality.

Results. The Georgia trial court (the original court) found no reason for the test and did not release the money. In their decision, they stated, "The theory of a genetic connection is not at the level of scientific acceptance that would justify the release of the funds."

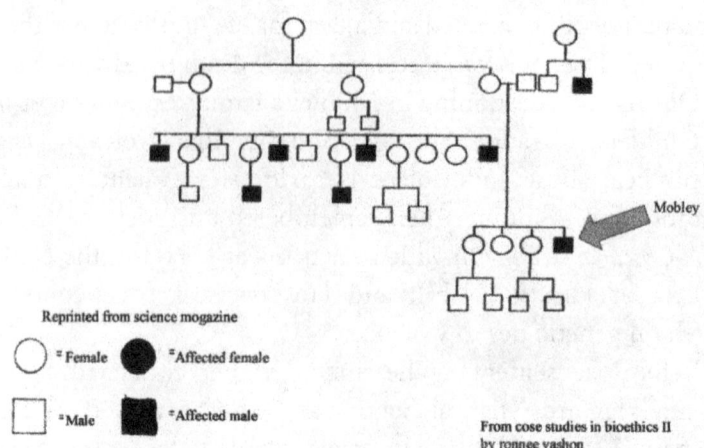

Illustration 6.1 Pedigree of the Mobley family

CHAPTER 7

How Do We Study the Effects of Genetics on Behavior?

Well, if we could find two or more people with the exact same genetic code and compare their behaviors, we might find some differences.

But who has the exact same genetic code?

Wait, identical twins do.

Nature versus nurture might be studied among twins separated at birth, and it has been.

Many studies analyzed individual questions such as twins with different fathers, intelligence quotient (IQ), obesity, schizophrenia, homosexuality, and fingerprints.

CHAPTER 8

Twin Studies

Twin studies have been going on for many years; even then, we had no clue about how complex twins can be. Many scientists made observations, and so did the parents of twins. As you already figured out, there are two types of twins: monozygotic (identical) twins and heterozygotic twins (two different egg and sperm fuse). We talked about what a perfect scientific study they would be. They might teach us something.

Note. With the development of assisted reproduction, many more multiple births (triplets, quads, quintuplets, and more) have happened because when an embryo is produced and the mother wants to implant (two or more frozen embryos), the doctor usually inserts a number of embryos (1–5), and then, they wait.

Twins reared-apart studies have certainly helped generate some interesting hypotheses, but they have completely failed to provide scientifically acceptable evidence in support of genetic influences on human behavioral differences, which include an intelligence quotient (IQ), personality, and psychiatric disorders.

One interesting case (the Dionne Quintuplets):

- Born in the backwoods of Canada
- A midwife thought they would not live and placed them near the fire
- A doctor came to see them and said, "I will raise them."

Illustration 8.1. The most famous identical quintuplets: Dionne quints

For twenty-five years, they lived with him and his nurses. He built a house with a large window all around and charged people who come and watch them play, eat, and open presents.

They never saw their parents until they were teens.

Important studies

The Two Jims (famous case in all biology books)

In 1979, Jim Springer and Jim Lewis, *the Jim twins*, were reunited at age thirty-nine after not knowing the other existed. As described in Segal's book on the identical Jim twins, *Born Together—Reared Apart*, both had been adopted and raised by different families in Ohio, just forty miles apart from each other. Despite their separate upbringings, it turned out that both twins got terrible migraines, bit their nails, smoked Salem cigarettes, drove light-blue Chevrolets, did poorly in spelling and math, and had worked at McDonald's and as part-time deputy sheriffs. But the weirdest part was that one of the Jim twins had named his first son James Alan. The other had named his first son James Allan. Both had named their pet dogs Toy. Both had also married women named Linda; then, they got divorced, and both married women named Betty.

Is this possible?

A few other cases

Two girls adopted in China by two American families did not find each other until high school. They lived in separate cities but were bought together by a friend who was shopping and saw a twin of her friend.

Later, they traveled together to China and met their birth mother.

"There are some twins in Brazil where one twin has microcephaly due to the Zika virus infection and the other does not," he adds. "And you'd really think, 'Hold on a minute, how does that happen if the mother gets infected, why does only one twin get infected?'" As a recent study on twins exposed to Zika in pregnancy suggested, infection risk could be related to epigenetic mechanisms.

Two of the identical twins raised apart (Jorge and William) shared the same bump on the same spot on the bridge of their nose, and until they were reunited, both had been convinced that it was from an injury. They also both preferred only eating the drumsticks of chicken. But one wore glasses, and the other did not. The other identical pair (Carlos and Wilber) had both been smokers, and both had a speech impediment.

CHAPTER 9

Legal and Ethical Questions

In the topic of this book, we run into a kind of collision (legal actions, divorce, murder, and of course, law), and ethics are separate topics; but because this not only is *nature v. nurture*, the discussion of the two is a mashup and shows up in some odd areas. How should we raise our children? Is there a difference in the way we, as parents, act because of how grew up or our genetics? This messes with legal verdicts in a court of law (some serious, some not). How do these questions sound to you? *Are parents libeled for what their children do? If a person has committed a serious crime and is a teen, should they be punished?*

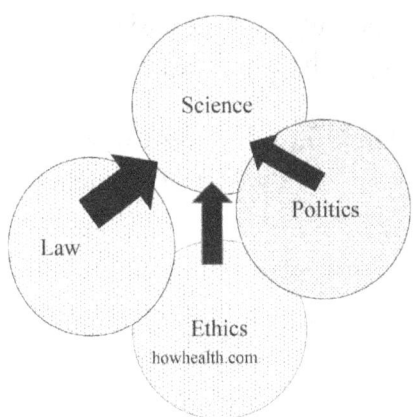

Illustration 9.1 Mashup

Example: This year, we have been exposed to a few examples. One of the most interesting is a parent or child mashup—the college admission scandal. Parents (well-known people) paid large amounts to get their children into college.

Human behavior multiple-choice question related to the preceding case

  a. The parents should be punished.
  b. Only the parents and the universities should be punished.
  c. The children should be punished.
  d. None of the above
  e. A and B
  f. *Only* A

Answers

  a. The parents got jail time and huge fines, but is that enough?
  b. Universities must have people spearheading this kind of scam.
  c. Did the children have a part in this?
  d. *No one?* Why could this not happen?

Note. The children are *not* enrolled at this time. Did the universities drop them because of the scandal? Should they be punished?

Table 9.1

Case	Ethical question	Genetics	Law	Outcome
University scandal	Why did the parents do this?	Were the parents assuming their child would not get in?	Were the children part of this?	Parents held responsible
As a teen, a child steals from a store.	Is this the parent's fault?	Why did they steal?	Who was harmed?	Maybe no one?
As an adult, the child commits a larger crime	Serious ones	Possibly, but even testing how to prove it	But are we ready to accept such punishment?	Depends on the age
Are parents *ever* responsible?	Their influence?	There have been some cases?	This is nature versus nurture.*	Example: School shootings

CHAPTER 10

Famous or Infamous Cases

In the law and in actual life, we judge a person's behavior with our own moral code. But in a court of law, there are more restrictions and controversy.

Each of these cases has been about how our brain and its memories and genetics apply to the reason for behaviors. As the scientific study of DNA, genes, brain chemicals, and age (brain size) continues to gather information, it is an application that can be applied in actual cases.

Table 10.1

Table 10.1	Name of case	Type of crime		Actual question	Background
	Frye v. USA	Use of lie detector as evidence		Should lie detector results be used as evidence?	

1	*Hinkley v. USA* (October 09, 1980). See chapter	Assassination attempt of Ronald Reagan and wounded James Brady	Found not guilty by reason of insanity and spent many years in a mental hospital	Was Hinkley insane? Eventually, he was released to live with his mother in 2018.	Because he was in love with Jodie Foster, he shot at Reagan to get her attention; movie connection
2	Andrea Yates	Drowned all her children in the bathtub	She had just given birth.	Is postpartum depression insanity?	Many thought no one could be sane and do this.
3	*Roper v. Simmons*	With three of his friends, they pushed a woman from a railroad trestle.	Simmons was sixteen.	Should a juvenile be tried as an adult?	Brain chemicals? Brain damage? Just a teenager?
4	Richard Speck	Killed three student nurses in their apartment, but left a witness	I lived in Chicago, watched TV showing the police, carried body bags out.	Was he mentally ill?	He killed all, but one, who hid under the bed and escaped. Later, she testified.
5	John Wayne Gacy	Murdered many boys in Chicago	Picked them up, killed them, and buried them in his basement	He and his wife lived in the home until he was caught.	He worked at children's parties dressed as a clown.
	Other legal theories based on behavior				

	Battered wife syndrome	For years, women who were beaten by their husbands were blamed if they fought back.	First case		
	Use of steroids in athlete violence	Think about the number of athletes who harmed or killed their wives.	Cases: *People vs. Simpson*		

Interesting Addendum

As you have seen, it is not clear what wins the *nature versus nurture* argument, but think about this: Since the beginning of the era of assisted reproduction, things have changed quite a bit. Problems with infertility in both men and women have brought to light the following:

- Sperm counts in men are decreasing.
- Women are waiting longer to have children.
- How this research changed my thinking.

What does it seem we need? Sperm and egg donors.

In the past, we have picked out mates by geography* (not too much travel); but as all society changed, we could not find mates.

So clinics began to cater to men and women who were infertile and began collecting samples of sperm and eggs stored in the frozen gametes and embryos to help. *But this was* not free!

Which sperm or egg donor would you choose?

Sperm and egg *banks* took long family and genetic histories (diseases, personality, criminal record as well as beauty, eye color, and intelligence).

Some sperm banks even offer look-alikes (Brad Pitt).

People would want what they want!

Remember the mashup: Can this affect what kind of child they create and, if we look ahead, narrow our choices?

Bibliography

Cummings, M. R. 2017. *Human Heredity*. Contributing Writers, Genetic Literacy Project Jack El-Hai. 2016. *XYY Men*.

Elysium, H. September 13, 2018. "The Science of Twins and How it Settled the Nature v. Nurture Debate." Endpoints.elysium-health.com.

Erika, H. May 15, 2018. *Identical Twins Hint at Environments Change Gene Expression*. The Atlantic.

Laura, F. July 21, 2017. "What Is Psychoneuroplasticity?" Originsrecovery.com.

Nancy, S. 2018. "Born Together, Reared Apart." Indiebound.org

Tim, W. (director). 2018. "Three Identical Strangers." *Documentary*.

Yashon, R. June 18, 2012. *Landmark Legal Cases in Science*, 5th ed. 2014 *RJ Publications*.

Yashon, R., and M. R. Cummings. 2015. *Human Genetics and Society, 2nd ed.* Cengage Learning.

Human Genetics
Its Influence on Behavior and the Immune System

PART B: IMMUNITY

Ronnee Yashon

Key Words

Immune
Bacteria
Inflammation
Lymphocytes
Antibodies
Antigens
Vaccinations
Blood types
AB
BB
AA
AO
BO
OO
Rejection
Transplantation
Cadaver
Living donor
HLA
Cirrhosis
Autoimmune
Immunodeficiency
Allergies
Anaphylaxis
Immunogenetics

Contents

Section B

Immunity

Preface

We are always hearing about new and exciting happenings in the science of human genetics. This book will be addressing just a few: immunity and behavior.

Illness is a part of life. For many years, we have known about diseases. But we were not always so smart. The black plague ravaged Europe between 1347 and 1352 and killed 30 to 80 *million* people. No one knew how people caught the condition. People just died. But not everyone died; possibly their immune systems helped them fight off these *invaders*.

Recently, the question about viruses have been on our minds. We see the name COVID and worry how we can all be seriously affected by it. Years of discoveries and vaccine development have allowed some more commonly known viruses (polio or flu) to be controlled and irradicated. Will this happen with COVID?

This section will explain how science is advancing but the public still questions science.

Today, we know about bacteria, viruses, cancers, and other invaders, and we usually let them run their course. But our bodies detect and fight them. That is immunity.

General Information

Most of us do not think about the immune system. But it works defending us and keeping viruses, bacteria, and pollution at bay. History shows we have amazingly improved. From the times of the black plague (see Figure 1.1) where doctors wore masks to keep from getting sick, today, we have medications that boost and slow down our immune systems.

Figure 1.1 Plague mask worn by doctors during the black plague

Doctors thought the shape of the mask would keep the *germs* from entering their bodies.

We will see how we cured (or at least stopped people from dying from) a virus called human immunodeficiency virus (HIV) and wiped out polio (a threat to children all over the world). But these are just a few.

In the beginning, B cells and T cells develop as white blood cells in the bone marrow. They both are lymphocytes. While B cells mature in the bone marrow, T cells travel through the bloodstream to the thymus gland—a small organ between the lungs and behind the sternum—and mature there.

This could explain where early in the diagnosis of leukemia, samples of the bone marrow are taken from the sternum.

When your doctor takes a blood sample for testing, you may wonder what he or she is looking for (see Table 1.1).

Table 1.1 Components of blood

	Purpose	Normal	Abnormal
Number of red blood cells	Carry oxygen	4.1–5.1 million	Low: slow healing
Number of white blood cells	Fight invaders (T and B cells)	4500–10,000	High: active infection Low: slow healing
Amount of hemoglobin	See how much oxygen is carried	14–17 g/dc	Oxyhemoglobin If low, oxygen does not reach the cells.

If the number of red blood cells is low, serious anemia can cause a lack of oxygen and is treated by iron supplements or transfusions if necessary.

If the number of white blood cells is high, it signals that there is an inflammation (infection). Or if very high, it might indicate leukemia.

If the number is low, it might need a study of lung function to see if the problem is there.

CHAPTER 1

How Does Immunity Work?

The immune system has three levels of protection.

First, the *skin* is a barrier to bacteria and viruses. Its cells are called epithelial cells. They are the outside layer of the skin (if a cut occurs, invaders can enter).

Second, an *inflammatory response* to the invader triggers the killer and helper T cells, and killer T cells destroy the infected cells. They recognize the infected ones by the molecules on the cells called antigens. The cells are marked, and the killer T cells destroy them.

Healthy cells have *self-antigens* on the surface of their membranes. They let the T cells know that they are not intruders. If a cell is infected with a virus, it has pieces of virus antigens on its surface. This is a signal for the killer T cell that lets it know this is a cell that must be destroyed. When the killer T cell fits the antigen receptor, poison is released (cytotoxin).

Third is the *adaptive immune system (specific targeted responses)*. It contains cells specifically made to identify and destroy foreign cells. Antibodies trap the invading viruses or bacteria in large clumps. This makes it easy for macrophages to eat them. Antibody-coated viruses are called *neutralized* because they cannot infect your cells.

Even after you have fought off your infection, some antibodies stay in your blood. If that virus tries to infect you again, your immune system has a head start trapping it.

Toward the end of each battle to stop an infection, some *T cells* and *B cells* turn into memory T cells and memory B cells. As one would expect from their name, these cells remember the virus or bacteria they just fought. They live in the body for a long time, even after all the viruses from the first infection have been destroyed. They stay in the ready mode to quickly recognize and attack any returning virus or bacteria.

Once an invader enters the body, *protein* molecules on the invader cell (called antigens) *activate the immune system.*

First, the detection of an antigen stimulates the T4 helper cell, which then activates the B cells.

The activated B cells then divide and produce proteins called *antibodies.*

Once the B cells are activated and released, they bind to the antigen wherever it is found in the body. Binding marks the cell for destruction by other cells of the immune system.

Another interesting fact is that some activated B cells become memory cells and can start a massive response if the antigen enters the body sometime in the future.

Note: Vaccinations are based on this fact. They include inactive (attenuated or weakened) forms of antigens and trigger an immune response.

CHAPTER 2

Blood Types

Before DNA, another form of blood typing was used in transplantation and other immune system problems.

Here is a little review:

The best-known blood typing system in humans is called ABO. The alleles (genes) that lead to any blood type come from the mother and the father—genes from the mother and half from the father.

A blood type is determined by these genes and how they combine. To understand this, refer to the following:

Blood type A has the alleles AO or AA.
Blood type B has the alleles BO or BB.
Blood type AB has the alleles AA and BB.
Blood type O has the allele OO.

Blood types do not change after birth and can be identified with a simple blood test.

You may be familiar with these blood tests in paternity testing or in your own medical checkup. When the question of paternity exists, blood tests are taken of the mother and baby and, if needed, matched with any possible father's blood.

Therefore, any of the following combinations, mentioned below in Table 3.1, can occur:

Table 3.1 Human Blood Types: possible genetics combinations

Mother	Father	Possible blood types
AA	BB	AB
AO	BB	AB or B
AB	OO	A or B
BO	BB	B only
BO	OO	B or O
OO	OO	O only

Blood types are used for determining whether samples are match types for blood transfusions and identification. If the wrong blood type is administered, the immune system will attack cells and cause a phenomenon called clumping (see Figure 3.1). If the wrong blood were transfused, it would cause a great deal of damage and possibly death.

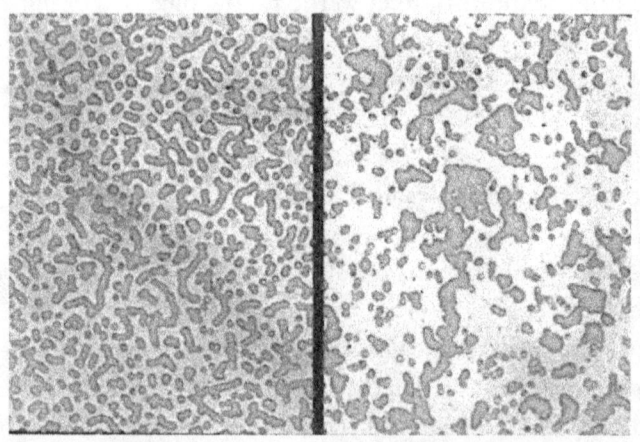

Figure 3.1 Blood test: (left normal, right clumping)

In this chapter, we are interested in how blood types are used as a preliminary test to begin finding a match for a transplant.

Before DNA, only blood typing was used in finding a donor. It was not perfect, but initially it could separate out the possible donors and non-donors. Family members were the first tested because they had similar genes as their siblings.

The immune system will reach out for any cells with a different antigen, not necessarily a disease. On a large scale, an entire organ (kidney and others) would make a perfect target for rejection. But many people have had transplantations and live healthy lives (see "Transplantation").

Although significant information is known about the immune system, we have not reached a time where organs are interchangeable.

CHAPTER 3

Transplantation

When an organ fails and is needed for life, it can sometimes be treated by medication (e.g., if the thyroid gland is removed, a synthesized thyroid hormone can be taken in a pill form). But as an option, a donor thyroid can be used.

Table 4.1 Organs used for transplant

Organ	Obtained from	Condition
Kidney**	Living donor or cadaver	Severe kidney disease.
Pancreas	Living donor or cadaver	Cancer
Lung*	Lobe or Living donor or cadaver	Lung cancer
Liver*	Lobe or Living donor or cadaver	Part of a healthy liver can regrow.
Heart/lung	Cadaver	Cadaver Car accident
Heart	Cadaver	Cadaver Car accident
Bones*	Cadaver	Funeral homes donate

Bone marrow*	Living donor or cadaver	Leukemia
Skin*	Living donor or cadaver	Burn victim
Cornea*	Living donor or cadaver	Eye disease or blindness

Living donors can only be used with certain organs.

The availability of actual organs for transplant varies all the time (we will look at this later in this chapter). If a match is made, then surgery must be done almost immediately after a donor dies.

Who Might Be an Organ Donor?

The perfect donor is a young, healthy, and dead individual. It sounds strange, but a perfect person who fits this description is a *motorcyclist*. In a crash, the head hits the pavement, and many riders do not wear helmets. Brain death is necessary for organs to be removed, and these serious motorcycle accidents usually involve the skull and brain. Even in such a situation, the family can refuse.

Let us look at some other problems associated with organ donation.

Think about what you might want and whom to tell about it.

Possible Cases and Their Possible Decisions

Scenario A: A child dies, and the parents will not allow transplantation of his organs. They are informed of the number of tiny newborn hearts to the size.

Scenario B: A young man dies in a motorcycle accident; the ER doctors determine he is brain-dead but place him on a respirator and look for next of kin.

Scenario C: A woman is brought to the ER and cannot breathe on her own. They attach her to a respirator and call her family. Her parents and husband come to the hospital. They do not agree.

Who can decide if organs should be donated? Next of kin? Who is next of kin? Spouse? Adult children?

If there is no next of kin or they do not agree? Judges?

ER doctors and hospitals do not especially condone this due to possible lawsuits.

What can fix this problem? Written permission from the patient? A will with signature? A conversation with an attorney (possibly)?

Table 4.2 Scenarios about transplantation and donating organs

		Decision 1	Decision 2	Another possibility
Scenario A	A child dies, and the parents will not allow transplantation of his or her heart. They are told these are the hardest organs to find because of their size.	Parents prevail. The fetal heart is removed and kept in stasis until a recipient becomes available (time is important).	Parents agree after a doctor tells them how many babies are in need of hearts at birth.	Before any decision is made, the story is released to the press. Often, a court might decide if it is right to remove the heart[1] under these circumstances.
Scenario B	A young man dies in a motorcycle accident, and his brain is not functioning when he is brought to the emergency room (ER).	After looking for next of kin, the hospital gives up. The organs are never harvested.	To avoid a lawsuit, the hospital takes a new step and asks a judge to decide.	They keep him alive for a while until a recipient is found.

---

[1] A case not long ago involved a patient with a few relatives who did not want to release his organs for transplant. No one could decide. The relatives searched his home and found a note stating what he wanted. The organs were harvested. *What if they did not find the note?*

Scenario C	A woman is brought to the ER and cannot breathe on her own. The doctors put her on a respirator and call the family: husband, mother, and father.	Her family rushes over to speak to the doctor. No one can agree. The husband says she told him not to use a ventilator. The parents say to keep her on the ventilator.	A judge is asked to do the decision-making. He or she decides that her husband had her consent (not written but verbal). The organs are removed.	
Scenario D	In the ER, a person dies, and no *next of kin* can be reached. But his organs are healthy.	The hospital has few patients who could donate. They could remove his organs with no one's consent.	The need for organs is dire (see the following statistics).	

Legal rulings: If a person is comatose and connected to a respirator or dead from an accident, the *next of kin* must decide. Some hospitals want a written form with signatures, while others just need a verbal consent.

Who is the next of kin? Parents (also for small children), spouse, and children (for elderly). Most states follow this format. Big problems can arise when relatives do not agree.

Why? Because doctors and hospitals will be sued if this is done without consent. This is a fine line because the organs may work and save the patient.

Why Do People Refuse Organ Donation?

1. They want their relative to be buried intact. Usually, a family or religious tradition causes this.
2. Cost (if not paid for by the donor's family).
3. They think that doctors are not trustworthy.

4. They know about donation and wonder how the organ will be obtained.
5. Could they ever repay the family of the donor?
6. They wonder if their relative might recover.

How Do We Overcome This Gap?

There is a *huge* gap between donors on the transplant list and people who need organs. Something should be done, and here are some suggestions that have been offered:

a. Give the doctors the right to harvest organs if there is no next of kin.
b. Allow payment for the organ.
c. Make it easier to sign up to be on the transplant list (advertisements).
d. Change the law so that we begin carrying donor cards *only* if we *want* to donate (called presumed consent). More details are covered next.
e. Presumed consent is already in place in Belgium, Spain, the United Kingdom (2020), and the Netherlands.
f. Allow donors to check their own family for matches.

Figure 4.1 Card issued to those who want to make it clear[2]

---

[2] In Hawaii, there are ads that run a lot, showing young people doing different things (surfing, picnicking, etc.). Each looks up and says, "Check the box." Next, the video explains the box on your driver's license.

What Can Cause a Patient to Need an Organ?

Table 4.3 What conditions can lead to a transplant?

Needed organ	Condition	Other information
Kidney	Chronic kidney disease, missing kidney, accident	Certain kidney diseases are inherited. Anyone can live with only one working kidney.
Pancreas	Cancer, accident	Often damaged in car accidents.
Lung	Lung cancer, chronic obstructive pulmonary disease (COPD)	Of course, smoking is a cause, but pollution and exposure to chemicals at work (asbestos) are serious.
Liver	Cirrhosis, alcoholism, liver cancer	Serious drinking and/or cirrhosis caused by liver cell death.
Heart	Serious heart attack, weak heart	Absolutely no living donor qualifies here, but if the transplant fails, a new heart must be found as soon as possible.
Bones	Amputee, infections in bone	These types of transplants are normally safe, but if the donor died of cancer, it is passed on.
Bone marrow	Leukemia and other blood diseases	A bone marrow with new blood-making cells is given to the recipient.
Skin	Burns, accident	Create new tissue in a lab or get some from the burn lab.
Cornea	Cornea missing because of accident	Replace.
Uterus	Can be used if not damaged	Babies have been born from a woman with a transplanted uterus.

How Are the Recipients Chosen?

Tissue match: Human leukocyte antigen (HLA) (see Figure 4.1), blood type, and the health of the recipient are considered. Close relatives are good donors; the best are identical twins.

When transplantation first began, doctors knew that organs were not compatible between people, but the knowledge of genetics was limited. Even before that, pioneers were looking into familial donors and found that transplants between siblings and parents worked better.

But medical professionals knew that identical twins were exact copies of each other because they came from one fertilized egg. So it made sense that their organs would be compatible, but it took a doctor with a reputation to try it: Dr. Joseph Murray.

Here is his story:

Richard Herrick received the first successful organ transplant in 1954. He entered Brigham and Women's Hospital in Boston suffering from kidney failure. His brother, Ronald, an identical twin, volunteered to donate a kidney to Richard. Ronald just wanted to help his brother. The night before the surgery, Richard wrote a note to his brother that read: "Get out of here and go home." Ronald wrote a reply: "I am here, and I am going to stay." The transplant was successful. And the surgeon, Dr. Joseph Murray, was awarded a Nobel Prize for this medical breakthrough.

Ronald Herrick went on to become a teacher, while Richard married and had two children. Unfortunately, eight years after the transplant, Richard died of an infection in the transplanted kidney. He was sixty-nine in 2010.

Figure 4.2 Herrick brothers after surgery

But today, we have much more information about the genetics of a transplant and drugs to calm the immune system after transplant—drugs that stop rejection and slow down the reaction of the immune system so it will not attack the organ.

Calming the immune system can be dangerous because we lose some of our defenses, and other invaders (colds, pneumonia, and cancer) can enter and not be detected.

Also, scientists have found a group of genes called the human leukocyte antigen complex. It is a group of genes found on chromosome 6 (see Figure 4.1).

We discussed blood types. For many years, blood typing was all we had. With the explosion of genetic information, now we can make it possible to actually map chromosome 6.

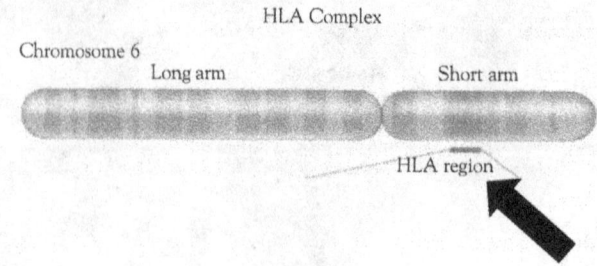

Figure 4.3 HLA complex on chromosome 6
Chromosome 6 showing where the HLA genes are carried.

Histocompatibility

The following things are taken into consideration (but not binding) when organs are transplanted. What organ is needed? How ill is the recipient? Is the donor healthy (living donors)? What is their age? What is the medical condition? And is it an HLA match (see Figure 4.1)?

Can doctors take your organ if it is needed for a seriously ill patient?
Can you sell your organs? Well, the answer is yes and no.

In most countries, it is illegal to sell organs or buy them. With the extreme shortage of organs (see the following), selling organs seems to be sensible, if it is kept to organs from living donors with their permission.

What are some of the drawbacks of putting organs up for sale?

Note: A number of years ago, a woman put her kidney on Craigslist. She did not know that this was not allowed by Craigslist. So they removed the ad. In the four days it was up online, she received fifteen emails.

Poor communities would benefit with funds paid but might be taken advantage of.

What are some of the advantages? As long as the donor understands everything, this could be done. And it has.

Does it violate any existing laws? In some countries, this is illegal; there is a black market.

Who benefits? (Donor? Recipient?) Everyone.

If the donor is already dead, should the relative get paid?

Which countries still allow payment? Israel and Holland are discussing this—China too.

The following chart shows some statistics (from the Organ Procurement and Transplant Network website, 2019) on organ donors and recipients.

Let us analyze these statistics. Yes, they are only numbers, but we can examine them closely and answer a few questions.

Table 4.4 Transplantation over the years

Year	Number of donors	Number of transplants	Number on the waiting list
2001	12,702	20,314	79,524
2002	12,821	21,523	80,790
2003	13,285	22,026	83,731
2004	14,154	23,266	87,146
2005	14,497	24,239	83,731
2006	14,750	24,910	87,146
2007	14,400	25,473	90,526
2008	14,631	27,040	94,441
2009	14,149	28,118	97,670
2010	14,011	28,940	100,775
2011	14,257	28,366	105,567
2012	14,412	28,964	110,375
2013	15,062	28,458	112,816
2014	15,947	28,662	117,040
2015	16,473	28,539	121,272
2016	17,554	29,534	119,364
2017	16,473	30,973	115,759
2018	17,554	36,529	113,759

As we look down columns 1 and 2, what do we see? As years changed, so did organ donations. Why? *Education of consumers? Advertising?*

As the number of transplants slowly increased, the number of patients on the waiting list shot up. Why?

The number of donors seemed to stay the same. *Better-informed doctors and patients?*

If only 13,821 people donate, why are there twice the number of transplants? *Can more organs be transplanted? Do recipients need more than one?*

What About Animal Donors?

(There is no written consent needed.)

Not all animals can be donors for humans. The closest are the great apes. But, as you might expect, the United States does not allow such transplants, except in clinical trials. But Food and Drug Administration (FDA) approval is needed.

We already have made some *big* forward steps:

a. Animal parts are used to substitute for a heart valve (porcine).
b. Animal tissue (porcine) is injected into the brains of Parkinson's patients.
c. The doctor will need permission to remove the organ and inform the recipient, if it fails, with an explanation about animal tissue compatibility.

Pig organs have other problems. Porcine virus may enter with the transplanted animal organ, and because the patient already has a compromised immune system, this virus might activate.

In some heart transplants, tests have been done with a new idea: chimeric immune transplant. In this method, the pig bone marrow is injected into the human bone marrow. When this happens, the recipient would not reject because the pig immunity would not trigger the rejection. However, this is still being studied.

A story about xenotransplantation: An interesting story occurred on October 14, 1984, when a baby (Baby Fae) was born with a hypoplastic left heart syndrome and needed a transplant to live. The problem was the shortage of tiny hearts, and Dr. Leonard Bailey had been working for years with apes. He spoke at length to Baby Fae's parents about using a baboon heart, which was the closest in size. When they gave permission, there was an uproar, but they stood their ground.

Years later, when asked for a copy of medical records, the hospital did not release them.

Results: Baby Fae died after successfully using the baboon's heart for twenty days. For some (Dr. Murray), this was very successful. But, to others, it was unthinkable to put your newborn (or a baboon) in such a situation.

Were these parents wrong?

Other examples of animal donors in use: Pig heart valves have been successful.

After a transplant, drugs will be administered in a difficult schedule for the rest of the recipient's life.

What if Rejection Occurs?

Because organs from one person might be attacked by the recipient's immune system, two things could be done to counteract the rejection:

1. The donor's organ should be as close as possible to a match or the recipient's immune system might be weakened (drugs or radiation). If this treatment was chosen, it would affect the body's fighting back for common conditions such as a cold.
2. The drugs used for this treatment have serious side effects and must be taken for life (see the following list).

Table 4.5 Antirejection drugs

Drug name	Side effects
Prednisone, steroid	Fluid retention, weight gain, fatigue, increased blood sugar, stomach irritation, irritability and increased alertness, and hunger
Tacrolimus (used in liver and heart transplants)	Abnormal dreams, agitation, frequent urination, general feeling of discomfort or illness, itching, skin rash, joint pain, loss of energy or weakness, and mental depression
Cyclosporine, antibiotic	Shaking, high blood pressure, infection, headache, nausea, and excessive hair growth
Mycophenolate mofetil (CellCept), antifungal	Nausea, swelling, rash, headache, and increased heart rate
Imuran (azathioprine), can increase chances of cancer, used with autoimmune	Nausea, vomiting, diarrhea, and hair loss
Rapamune (rapamycin, sirolimus), used after a kidney transplant	Tiredness, vision and hearing problems, weakness, and bone pain

CHAPTER 4

HIV and AIDS

The topic of human immunodeficiency virus (HIV) and acquired immune deficiency syndrome (AIDS) has been in the news for quite a long time. At one time (1980s), it was much more than a medical problem. It pulled in all the parts of our political and personal lives. In some areas of the United States, it also brought out prejudice and anger.

What Is the Difference Between HIV and AIDS?

HIV is a virus, also called human immunodeficiency virus. Specifically, this virus attacks the cells that make up the immune system. If untreated, it will eventually spread, causing AIDS (acquired immune deficiency syndrome) and death.

In the early years of HIV, doctors noticed rare conditions that they had never seen before: a form of pneumonia (*Pneumocystis carinii*) and a form of cancer (Kaposi's sarcoma). These are sometimes called AIDS-defining illnesses, because when the virus (HIV) attacks the immune system, it, in turn, *cannot* kill the virus (HIV) already in a cell and the cell dies. These diseases are extremely rare. If a patient has one, it alerts the physician to look for AIDS.

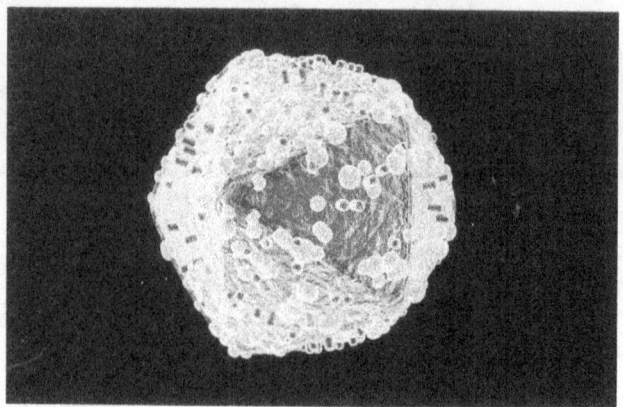

Figure 6.1 HIV

Human Immunodeficiency Virus, or HIV

All viruses work by reproducing themselves inside a healthy cell. Then, the cell breaks open, and all these viruses move throughout the body, and HIV is no exception. This also kills the T cells, and when the number of T cells is low enough, HIV becomes AIDS.

When the white blood cell (WBC) count drops down, certain diseases (opportunistic diseases[3]) and symptoms show up because the body's immune system is compromised. As the HIV-affected cells increase, even simple diseases become life-threatening (e.g., cold).

HIV has a long history before and after the treatment was available. The treatment was sped up because of organizations such as ACT UP (AIDS Coalition to Unleash Power) forcing the FDA to speed up the process of releasing the treatments.

It took a while, but after the treatment was released to doctors and then patients, they were *quickly* used and saved lives.

If not treated, over time, the WBC count drops lower and lower, and death occurs from other diseases.

---

[3] These are rare because only someone with a compromised immune system can have them. Some examples are *Pneumocystis* pneumonia, thrush, and Kaposi's sarcoma (a form of cancer).

How Does One Get HIV?

After much discussions and statistics taken at the time, it seemed there were a number of ways: sexual contact, blood transfusions, organ transplants, blood products, and any other contact with blood[4]—but *not* kissing, touching, or sharing food. This turned into a full-blown panic in the United States; people were making illogical choice, with no real knowledge of the condition. For example, many thought by *not* using dishes or cups from HIV-positive people, they would stay safe.

Read the case of Ryan White in the next chapter.

Problems with Diagnosis

At first, the United States had a few cases of a rare form of pneumonia (*Pneumocystis carinii*), but it began to show in certain groups. The way it was treated and diagnosed was a surprise because it was so rare. And so doctors began to discuss this and discovered that it had been found in certain groups, such as homosexuals, Haitians, children born from infected women, people who had had blood transfusions with HIV-infected blood, and hemophiliacs.

As this sometimes happens (as we all know), the truth has problems being explained, and panic occurs. Some examples were people getting fired from jobs or not hired, just because they were part of a group that was affected. It also was suggested that, if a patient tested positive for HIV, he/she should be confined to a specific area of the country. Read the case of the *devil*.

---

[4] A dentist who was HIV positive passed the virus to a number of patients, without telling them he was HIV positive. This was not only unethical but also illegal, and things just escalated from there.

Milestones in HIV

Case 1: Saving the Devil

Researchers in Australia came across a seemingly unrelated case.

Not so many years ago, a rare mammal was found on the island of Tasmania (near Australia). This island was small but had many rare species because of its isolation.

Enter the Tasmanian devil (yes, this is also a cartoon character).

Over the years, the population of Tasmanian devils dwindled. The cause seemed to be cancerous tumors growing on their snouts.

Australian scientists went to study the disease that was wiping out the Tasmanian devils (see Figure 7.1). Cancer? Really? How did they get this virulent, deadly disease? It was communicated from one to another.

Figure 7.1 Tasmanian devil

The study noticed the tumors were always on the side of the snout and the devils died quickly. During this time, the scientists watched them mating and saw an amazing thing. During sex, the male bit the female on the snout, almost exactly in the place where the tumors would grow.

The keepers wanted to isolate those with the cancer to stop the spread and the ultimate extinction of the devils.

What does this have to with the human immunodeficiency virus (HIV)?

On a much larger scale, HIV was killing people, certain groups of people. If those groups were put in an area by themselves, would this isolation possibly work and make others be safe? Didn't this work with leprosy?

When HIV first became an epidemic, it was found mostly in men. They were of four groups: homosexuals, Haitians, drug addicts, and people with hemophilia (an inherited blood condition).

The public was asked to come in for testing for HIV, but some people were afraid. And, as with other sexually transmitted diseases, they were told to give the names of their partners to the health department.

This law was important to finding the spread of HIV. But people still were afraid because they knew their jobs, their families, and their friends would not want them around. Some of these patients, who were tested, did lose their jobs and could not find any other. If a doctor tested a patient and he or she was positive for HIV, then they would not be hired and otherwise be discriminated against. Where was our right to privacy? Does it occur when serious diseases (another example, Ebola) threaten to spread quickly?

But how did the doctors, employers, and businesses find out the test results? Doctors could not give out this information. But, soon, they were required to by law, so the government could track the cases. Then, people were really frightened. But the law did contain one thing: the results were sent with no names, only numbers, and *no* medical records were available.

Committees in the Centers for Disease Control and Prevention (CDC) were asked to find partners on the list and inform them of their danger, offering testing and follow-up.

In Illinois, a law was passed that anyone applying for a marriage license must present proof of taking an HIV test, but the results were not on the medical record given to the state. So what about the two people marrying? Did they have to tell each other?

At this point, we come upon the most striking of the problems: *discrimination* on every level. This topic had been faced in many venues: age, race, religion, monetary means, and many others. We are all familiar how it could be applied in employment, marriage laws, and areas to live in—where we, humans, were never sure how others would look at us or treat us. Most of us have encountered this with issues such as race, religion, housing, money, or political beliefs.

Oddly enough, you did not even need an HIV diagnosis to be discriminated against.

Just to be in a group where HIV was prevalent was all you needed.

Remember the Tasmanian devil? Would this work for people?

Case 2: Ryan White (Discrimination)

Ryan White was an American teenager from Indiana, who became a national poster child for HIV/AIDS in the United States after failing to be readmitted to school following a diagnosis of AIDS. As a hemophiliac, he became infected with HIV from a contaminated factor VIII blood treatment and, when diagnosed in December 1984, was given six months to live. Healthy for most of his childhood, White became extremely ill with pneumonia in December 1984 and was given a lung biopsy on December 17, 1984. White was diagnosed with AIDS. AIDS was poorly understood by the general public at the time.

When White tried to return to school, many parents and teachers rallied against his attendance because of concerns of the disease spreading. In addition, when the state announced he could return, that day no students or teachers came.

Before Ryan White, AIDS was a disease stigmatized as an illness impacting the gay community, because it was first diagnosed among gay men. That perception shifted as Ryan and other prominent straight HIV-infected people—such as Magic Johnson (basketball), Arthur Ashe (tennis), and the Ray Brothers—appeared in the media to advocate for more AIDS research and public education to address the epidemic. The United States passed a major piece of AIDS legislation, the Ryan White CARE Act, shortly after Ryan's death. The act has been reauthorized twice.

Case 3: ACT UP (Grassroots Organization)

Getting Things Moving

The FDA (through a time-consuming and costly process) approves new drugs only after animal testing and four stages of human testing called clinical trials.

This process takes years and costs millions of dollars. So what happens if a condition is life-threatening and affects many people?

In the 1980s, this was a situation where life was threatened because of HIV. Thousands of people were dying of AIDS, but the cost and the time commitment slowed down everything.

Leaders of an organization called ACT UP began using social activism to generate change, bring more drugs to the market, and increase the awareness about HIV. This small organization of gay men and women organized a grassroots movement to get AIDS into the national spotlight. How do you get the attention of the public?

In June 1989, one of the most successful protests was held in front of the Sloan Kettering Hospital in New York. Protesters dressed in healthcare worker outfits and patient outfits. They sat in front of the hospital for four days while acting out scenarios about people dying of AIDS.

Factor VIII is the blood component that helps the blood clot. Hemophilia has been known for many years because of the stories told about a Russian royal family (the Romanovs) whose only son, Alexei, had the condition. Of course, no one understood what it was, and he was desperately ill. He was the heir to the throne. Oddly though, he did not die from the condition because his family was overthrown and they all were killed in July 1918.

The Ray Brothers were three brothers diagnosed with HIV in 1986. The citizens of their small town burned their home.

The purpose of the protest was to demand that more people with HIV be included in the clinical trials for these drugs.

Because of the pressure brought about by ACT UP and other groups, the FDA was forced to make HIV drugs available for treatment under an already existing rule called *compassionate use* rule that allowed new drugs to be moved through the trials more quickly. This was the first step.

Finally, in 1997, the FDA formally introduced the fast-track designation and priority review.

Case 4: HIV in the Blood Supply (1980s)

It was known early on that HIV was a blood-borne disease, but for some reason, the medical community did not see this as a danger to the donated blood supply. People with HIV continued donating their blood without a thought (most did not know they were infected). After patients with HIV, who did not fit into known groups, were found to have the virus, the American Red Cross was alarmed and demanded testing of all blood that was donated. Lawsuits were filed for negligence, but they were dropped.

Case 5: Does the Government Have an Obligation to Protect Its Citizens?

According to the constitution of most states, they do. Not only that, if an epidemic enters our (such as HIV) country, what does protection mean? Place those with the virus in areas and give them everything they need.

Does it mean finding out who is infected and, if no treatment is available, setting out rules to follow? Would this take away people's rights?

Timeline of the Epidemic

- 1978: Early case of AIDS in the United States (baby born to a sixteen-year-old drug user).
- 1980: A case of Kaposi's sarcoma in San Francisco (first case in the United States).
- March 7, 1980: Headline in *NYT* read, "Rare Cancer Seen in 41 Gay Men."
- July 4, 1980: Clusters of opportunistic conditions reported.

- By the end of 1980: 121 men had died.
- 1982: First cases in Spain, the United Kingdom, Italy, Brazil, Canada, and Australia.
- 1983: A retrovirus identified as HTLV-I, and then HIV was the cause.
- 1983: Blood donor screening guidelines (including high-risk groups excluded from giving blood).
- 1983: PCR invented by Kary Mullis, used in AIDS research.
- 1984: Cases in Portugal, the Philippines, China, Soviet Union, and Italy.
- 1984: Robert Gallo isolated HIV.
- December 17, 1984: Ryan White diagnosed with AIDS.
- March 12, 1985: ELISA test for HIV in the blood was available.
- 1986: One million Americans had been infected with HIV.
- 1987: First antiretroviral (treatment for HIV) became available.
- 1987: ELISA test gave rise to the Western blot.
- March of 1987: ACT UP formed.
- 1987: First clinical trial of AIDS vaccine.
- 1989: Film about the life of Ryan White, called *The Ryan White Story*, came out.
- 1989: HIV was discovered in newborns.
- April 8, 1990: Ryan White died.
- 1990: Congress passed the Ryan White CARE Act, which allotted funds for HIV.
- 1996: HIV-resistant gene (CCR5-D32) found.
- 1996: Number of HIV patients was now 1.5 million.
- 2003: George Bush passed the Emergency Plan for AIDS Relief.

Scientists have found the cause of HIV, how it was passed from monkeys to humans (eating monkey meat), how it was passed from human to human, and, finally, treatment for it.

Now, HIV patients can live full lives and never move to having AIDS.

The only thing left now was finding a prophylactic that would work to keep healthy people from contracting it. Is this impossible?

Yes, but I have a story. As you read it, think about this: if we could stop HIV, how much would it cost?

Robert Grant had just finished medical school, and one of his professors asked him, "What will you do now?"

The answer was "HIV research."

The professor surprised him by saying, "Do something else. By the time you are finished, the epidemic will be over."

Grant thought very seriously about this statement. He knew that working on a cure or a vaccine had been unsuccessful, but he had another idea.

The antivirals have been successful in keeping the viral count, and even though you have to take it the rest of your life, you will not move into AIDS.

He raised grant money from the governmental and private funds to finance his plan. What was his plan?

To run clinical trials with people who had never had HIV. From the beginning, the study gave a drug to each of the participants, one pill per day.

The drug, *Truvada*, was an antiviral already used in treatment along with other drugs to stop the virus from multiplying.

Trials were run for eight years, using Truvada that was donated from its manufacturer. Each participant was tested for HIV and was found clear.

Then, every day, for six years, these participants went about their day taking one pill every day. Its name was PrEP (pre-exposure prophylaxis).

It worked! Not one of his participants on Truvada contracted HIV. This was more than he imagined, but the story is not over yet.

Numbers

One person on Truvada from age twenty-five to fifty would need one pill a day, 365 pills a year, and it turned out that the drug was under patent. The company donated pills for the trial but did not lower the cost to everyone.

Thousands of gay men, women, and drug users could use it without fear, but US $14,000 per year was impossible for most.

CHAPTER 5

Allergies

What Causes Allergies?

Allergies result from the immune system's reaction to allergens, which are carried by dust or certain foods. One of the most serious food allergies are allergies to peanuts. Allergies to peanuts, beestings, and other allergies have very serious reactions. This reaction is called anaphylactic shock. If a patient is not given medicine immediately, they will die quickly because of this reaction. Parents with children who have this serious allergy must have an EpiPen with them at all times. Eighty percent of all these serious allergy attacks are caused by peanuts. Children and adults with this allergy can die very quickly. Called anaphylaxis, it can occur during the *first* exposure. Symptoms include swelling of the tongue and throat, constriction of the airway, and drop in blood pressure.

This is so serious because only small amounts of the allergen need to be present for the reaction to occur. Some schools do not allow anything that contains peanuts (even dust given off by the peanuts themselves).

Peanut allergies have increased since 2008: 1 percent of all children in the United States have these allergies to peanuts. People in some countries (China) who do not eat many peanuts do not have these allergies. But in the United States, children are exposed not only from what they eat but from their mothers' breast milk. The reason for this seems to be twofold, and infants are very sensitive to these allergens. Doctors suggest that mothers not yet of their children have peanuts until they are older.

If you watch carefully, you may see signs in restaurants and other public places stating to not bring peanuts (Figure 8.1).

Figure 8.1 Peanut-free sign in a restaurant

Schools also have a problem if a child has a serious peanut allergy; they cannot be around any peanuts, so the schools must tell parents to not send peanut butter and other foods with their children.

Just like autoimmune diseases, allergies are triggered by inflammation caused by foods, chemicals, air quality, and formaldehyde (in new clothing and furniture). But allergies to peanuts are one of the most common in children and adults. It is estimated that 284.8 billion have allergies.

An interesting case: A baby was born with a serious allergy to cow milk, and it was identified early. The mother, who was proud of her baby, took her to see a friend. When the friend kissed the baby after taking a sip of coffee with cream, the baby's cheek swelled up in the shape of her friend's lips.

CHAPTER 6

Autoimmune Diseases

In previous chapters, we have looked at how the immune system protects us by identifying foreign substances, viruses, and bacteria, targeting them for destruction. But what if a donor organ is inserted in the body with foreign cells?

The same exact thing! You are probably thinking this can be tackled (see "Transplantation"). (But *why* does the immune system begin to attack *our own cells*?)

The immune system is controlled by several groups of genes. The study of these genes and how they function is called immunogenetics. They encode for proteins on the cell surface as well as the antibodies that directly attack foreign antigens. The receptors on the surface of our T and B cells are also encoded in our DNA. Understanding how these genes and their respective proteins work has helped us cure infectious diseases, prevent infection, and make organ transplants possible.

Because these genes control the immune response, mutations in these genes can also cause diseases of the immune system, including autoimmune disorders and allergies.

Diagnosis of autoimmune conditions has been difficult in the past; there was often no way to diagnose. But now there are treatments, along with knowledge about the DNA of viruses and bacteria. In addition, they can be overcome with steroids. This treatment works temporarily because of the toxic side effects of the long-term use of steroids.

Treatment here is similar to other conditions, such as human immunodeficiency virus (HIV) (see "HIV and AIDS") and acquired immune deficiency syndrome (AIDS).

In autoimmune disorders, listed next, the body fails to recognize its own cells and attacks and destroys them (this includes juvenile diabetes, arthritis, multiple sclerosis [MS], and inflammatory bowel disease). See Table 5.1.

When you look at this list, you are probably familiar with some of the listed disorders. Conditions marked with a * had a questionable diagnosis, but as we have learned more about autoimmune condition, more scientists are working on it.

Table 5.1 List of autoimmune diseases

	Area affected	Symptoms	Treatment
Allergies	Skin, other organs	Cough, breathing problems, asthma	Antihistamine, decongestant
Rheumatoid arthritis	Central nervous system joints	Pain, additional symptoms from other inflammation areas	Steroids, pain killers
Psoriasis*	Skin	Red and scaly patches	Steroids, creams, immunosuppressants
Narcolepsy*	Sleep center of the brain	Falling asleep many times a day	Drugs to stay awake, caffeine, psychological
Diabetes (type 1)	Pancreas	Very high and low blood sugars	Insulin, diet, regimented schedule
MS	Central nervous system	Begins in your twenties	Steroids, specific immunosuppression drugs
Graves' disease	Thyroid gland	Increase in thyroid hormones	Removal of the thyroid
Celiac disease*	Digestive system	Diarrhea, cramps, bloating, mouth sores	Gluten-free diet (to mitigate symptoms, check drugs to control symptoms)
Crohn's disease	Digestive system	Same	Same

Lupus* erythematosus	Systemic: bones, skin, and kidneys	Joint pain, weakness, hypersensitivity to light	Steroids (treat symptoms)
Alopecia*	Hair follicles	Clumps of hair falling out	Steroids (hair follicles die, and a wig is a must)
Pemphigoid	Skin	Hardening of the skin, liver involvement	Steroids, anti-immune
Scleroderma	Skin	Same	Same
Fibromyalgia*	Pain in the endings	Pain all over the body	Controlling pain, rest, lowering stress
Guillain–Barre syndrome	Face and other parts	Partial paralysis of the face and other parts	Muscle relaxants, stress relief (not much treatment, but clears up by itself)
Restless leg syndrome*	Legs	Movement of the legs while sleeping, hard-to-stop pain when awake	Mild sleeping pill

Do you know anyone who has one of these conditions?

The Autoimmune Spectrum

The symptoms of an autoimmunity can continually become worse without treatment of some kind. The progression of the illness can be calculated by using the following illustration:

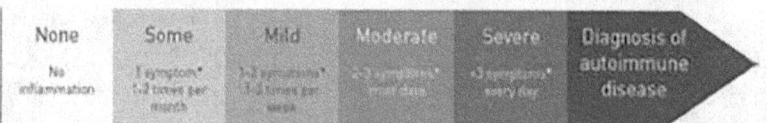

Figure 5.1 Diagnosis of autoimmune conditions

As time goes on, more symptoms surface and increase in intensity.

CHAPTER 7

Legal and Ethical Questions

Each of the topics we have covered has serious ethical questions that people (possibly you or your family) might encounter.

Let us look at some of these and how the law has addressed them:

Transplantation:

	Legal solution	Problem	Causes	Other
Donation	With permission, everyone.	Low donation rate (see "Transplantation").	Lack of understanding.	
When is the organ removed? When is death?	Time decided by the patient's illness. Malpractice suits occur.	At the time of death.	No one really knows when this is.	Heartbeat stops? Breathing stops? Brain death?
Who gets an organ?	The law sets guidelines.	Someone might be pushed to #1.	Money, influence, celebrity.	

Autoimmune conditions:

Where do they come from?	Diagnosis is difficult.	Another problem.		Treatment

Family genetics	Many doctors have little knowledge.	The basic symptom involves the immune system.	Large-scale symptoms make it hard to treat.	Uses immune-suppressant drugs with many side effects.
How to diagnose?	Specialized testing.	Problem.	Another problem.	
Can anyone get more than one?	Expensive and probably not well understood.	Doctors often identify them as *stress related*.		

Allergies:

Question	When does it occur?	Problem	How to treat?	
Is this partially genetic?	Newborns to adults.	Too many over-the-counter drugs available	Depends on symptoms (stuffed nose vs. breathing stops).	Runs in the family (e.g., hay fever).
Fatal?	It could be.	Peanuts	Keep away from peanuts.	Carry an EpiPen.
Are they real?	Some say they are in the mind.	This has been debunked.	Some of these diseases are long-term and will not respond to medicines.	The symptoms can come and go.
	Specialists called allergists can treat with allergy shots.			

HIV/AIDS:

Question:			
What has happened?	Safe sex and antiviral treatments give a person a much longer life.	Many famous celebrities came forward and urged people to get tested.	Although a virus, it has needed a lot of time to allow its treatments in the market (see Appendix A).	
Blood transfusions	At first, people with HIV were donating blood.	Then, certain groups could not donate.	But later, a test of the donated blood came available.	Still, many donors are refused.
Treatments	They are complex (antiviral and lifelong regimen).	Constant checking.	Still, the virus is mutating, so it is not over.	
Nonmedical problems	Homosexuals were horribly treated even if they were clear.	Many people were not *hired* because they were *suspected* of being gay.	AIDS medication is amazingly expensive. Not available there.	Many wonderful and creative people died from this.

CHAPTER 8

Interesting Questions from Issues in This Topic

These little charts ask you to think of some of the scenarios this book presents and put yourself there. Maybe it will help to see my opinions and why I chose them.

Transplantation:

What would you do if…	You	Me	My reason
You needed a kidney?		Look for a living donor (organ donor website).	It would depend on the situation and person.
You wanted to talk to your family about donation after death?		There are forms called informed consent. I have already done this.[5]	Just say, "Let us look at this form." And read it together. Talk a lot.

---

[5] I was very affected by a case where a young woman, engaged to be married, was in a car accident and needed a respirator to breathe. Her parents and her fiancé did not feel the same way. The parents thought she should stay on the respirator because they knew she would wake up. But the fiancé said that he was told, *by her*, that she did not want to be on a respirator. But no written document existed. Two groups picketed the hospital: church groups and people who felt she had a right to a death of her choosing. The court found for the fiancé.

You wanted to donate your organs after death?		*Yes*, without hesitation (I indicate such on my driver's license).	It is a wonderful way to help people and the environment.
Would you take an HIV test if asked to by a friend?		Well, when I got married, Illinois was the only state to mandate an HIV test for a marriage license.	Oddly enough, no results had to be shown, just the form signed by a doctor stating the test was given.[6]
Would you ask a friend if they were HIV positive?		This is hard. I am not sure. It does not seem to be my business.	This would be something you could ask a *significant* other.
Would you tell a person about the importance of vaccines?		Give them an article to read and point out in simple terms why this cannot be connected with another disease.	Hard to do this because of the lack of basic science education. "How Does the Immune System Work?" might work.
You were asked to donate bone marrow to a child with leukemia?		Yes, but it would depend who the child was.	Bone marrow extraction is not really that bad.

---

[6] Note: The HIV test cost US $100 each, and the marriage license was US $5. We were both negative.

About the Author

Ronnee Yashon is a nationally known expert in teaching genetics, ethics, and the intertwining of law on all levels. She has a background in teaching and speaking in the high school, undergraduate, graduate and law school levels. Her case study methodology for introducing bioethics and law uses simple, personalized, and current scenarios that involve the students in decision-making.

Book Awards

- Presidential Award for Outstanding Science Teaching, 1992
- Finalist for the H. Dudley Wright Foundation Teaching Award, 1992
- Awarded Problem Based Learning Grant of $3500 from Hitachi Foundation, 1991-2
- Nation winner, Tandy Technology Scholars, awarded by Tandy Corporation, 1991
- Christa McAuliffe award for outstanding Science Teaching, 1990

Current Publications

Author, Human Genetics and Society Collection.
    Series of 10 (see list below)
    Published by Momentum Press. 2017- present
    List of Titles:

- Fertility. Infertility and Treatment Options
- Published 2017
- Chromosomes
- Published 2017
- Genetic Testing
- Published 2018
- DNA Forensics
- Published 2019
- Biotechnology
- Published 2019
- Genetics and Human Behavior.
- Published 2021
- Human Genetics and the Immune System
- Published 2020

Printed in the USA
CPSIA information can be obtained
at www.ICGtesting.com
CBHW020350041024
15321CB00056B/1835